Seedbed to Harvest

Farm Machinery Through the Seasons

62
FL7
STAMMERS
66
E R F
M265 DCL
3

Seedbed to Harvest
Farm Machinery Through the Seasons

by

Jonathan Whitlam and
Stephen Richmond

Farming Press

AUTHORS' NOTE

All figures (such as horsepower and capacities, lengths, etc) given in this book are approximate and should be used as a rough guide only. Machine specifications are constantly changing and being updated; equipment fitted to the machines featured may not be included on every machine of that type.

We would like to thank James Armstrong for kindly providing the photographs on pages 12 (top), 29 and 57 (bottom). All the other photographs are from our own collection. We would like to express our gratitude to all the farmers, contractors and farm workers who made these pictures possible; there have been far too many over the last fifteen years to mention personally.

Front cover

Two Ford 8340 tractors prepare land for potato planting with Reekie de-stoners.

Frontispiece

Pea lorries wait on the headland as two FMC 979 pea harvesters approach.

Back cover

Top left: A Mengele Mammut self-propelled forage harvester makes short work of a forage maize crop with a Massey Ferguson 6180 and trailer alongside.

Bottom left: Autumn ploughing with a Ford 8630 and five-furrow reversible plough.

First published 2000

1 3 5 7 9 10 8 6 4 2

ISBN 0 85236 534 9

A catalogue record for this book is available from the British Library

Published by Farming Press
Miller Freeman UK Ltd
Miller Freeman House, Sovereign Way
Tonbridge, TN9 1RW
United Kingdom

Distributed in North America
by Diamond Farm Enterprises
Box 537, Bailey Settlement Road
Alexandria Bay, NY 13607, USA

Origination by TSS Digital, Margate
Printed and bound in Great Britain by
Stephenson George of Merthyr Tydfil

PREFACE

In this book we travel through the four seasons, taking a look at the various tasks on the farm, the equipment used to perform them, and the different techniques employed. We start in the spring with the preparation of the land ready for spring-sown crops – such as barley, sugar beet and potatoes – before moving on to the start of the silage harvest. Summer arrives and we are into pea harvesting and the all-important grain harvest, before moving on to autumn and the hectic rush to prepare the ground for sowing wheat and barley, as well as lifting sugar beet and potatoes. Winter comes last and proves to be surprisingly busy, given the correct weather conditions.

Of course, because almost all the photographs in the book are from our own collection we have not been able to include every farming operation or indeed anywhere near every piece of farm equipment, but we hope that what we have included will give the reader a good idea of what takes place throughout the farming year as we leave the twentieth century.

In *Seedbed to Harvest* we have concentrated on the farm machinery used in a variety of operations, as well as how it operates. You can find out more about tractors or self-propelled machines in our two previous books: *Modern Tractors* and *Modern Farm Machinery*. The intention with the present volume is to put the machinery featured in the earlier books into the broader context of each farming operation.

Unlike other farm machinery books, this volume – and the others before it – feature mainly our own photography. We prefer to show tractors and farm machinery working in real conditions rather than brand new, or restored machines, posed in brilliant sunshine or photographs taken from glossy sales brochures. We believe that the only really interesting way to show how modern machinery actually makes a difference to everyday farming is to show it warts and all – with dirt and mud covering the paint work and the weather doing its best to ruin everything.

Farming changes constantly and has certainly done so for the last hundred years or more. Whatever the future holds for agriculture we hope that this book will stand as a pictorial record of what farming was like at the dawn of the third millennium. So sit back and enjoy a journey through the modern farming year, from seedbed to harvest and back to seedbed once again.

JONATHAN WHITLAM
STEPHEN RICHMOND
January 2000

INTRODUCTION

The farming year is governed by the weather and in spite of the appearance of the high-tech machinery that the modern farmer takes for granted, this statement is as true today as it was a hundred years ago. That said, the huge variety of machines available today means that, short of being able to control the weather itself, the farmer is now more in charge than ever before.

The internal-combustion-engine tractor was what really started off large-scale agricultural mechanisation from the 1920s. It proved to be much more reliable and cheaper to run than the steam-powered traction engines still popular at that time. British farmers however were slow to adopt what they saw as a new-fangled piece of equipment.

With the coming of the tractor, more and more manufacturers saw how this versatile machine could be used to carry out a great variety of farming tasks. They began designing and building equipment to be used with it, making full use of the tractor's drawbar, belt-pulley, power-take-off shaft, hydraulics and eventually the three-point linkage, as these gradually became available. The driving force behind the development of all forms of agricultural machinery has been to reduce the amount of labour needed to carry out routine farming jobs. This process has seen the countryside change as traditional methods of soil and crop husbandry have evolved to keep pace with modern technology.

For centuries the horse was the only form of power on the farm. Many breeds were developed specifically for the demanding tasks of agriculture. One such breed was the Suffolk horse, shown here pulling a single-furrow plough. The ploughman would have to walk a great many miles each day behind his team, but his job was seen as the most prestigious and important on the farm.

As an example of just how much tasks have changed, take the cultivation and harvesting of sugar beet. This root crop did not catch on with the British farmer until the start of the Second World War, when supplies of sugar cane from warmer countries were in short supply. Sugar beet was a very labour intensive crop. The space between the rows of young plants needed to be kept free of competing weeds, the crop also needed thinning and all this had to be done by hand with a hoe. When the crop was ready for harvesting the roots had to be pulled from the ground, topped and then loaded into carts to be taken away from the field, all of which again was carried out by hand. Today the precision drill has made thinning a thing of the past, the tractor-mounted hoe and modern herbicides have replaced the hand hoe and the huge variety of machinery capable of harvesting from between one to nine rows at a time have replaced the back-breaking toil of lifting the crop.

The self-binder replaced the reaper as the harvesting tool – which had in turn replaced the hand-sickle – it also tied the crop into sheaves. A David Brown 990 is seen at work with a Massey Harris binder in a crop of wheat. Two men were needed, one to drive the tractor and the other to operate the binder from a seat at the rear. ⏷

⏶ *In the 1940s the Ferguson TE20 revolutionised farming with its three-point linkage and hydraulic depth control. The plough could now be mounted on the tractor and lifted at the headland. This TE20 is seen with a two-furrow one-way plough. Modern tractors still use a three-point linkage based on the original Ferguson System.*

Mechanisation has changed the landscape and put many thousands of farmworkers out of work. The demand for cheaper food has forced farmers to run their holdings as efficiently as possible, the result being a need to invest in larger and more sophisticated machinery.

One of the latest innovations being adopted by an increasing number of growers is precision farming. This technique makes use of yield maps generated by computer when the cereal harvest is under way. A global positioning system in the combine cab links up with satellites, enabling the on-board computer to work out its position and the yield being obtained in each part of the field. The yield maps are used for all subsequent operations in that field, from seed rates to spraying and fertilising rates This means greater efficiency in the use of seed and chemicals, as well as saving money and time. GPS will gradually become part of the farmer's basic equipment and one day it may be one of the building blocks for driverless tractors and harvesters. Anything is possible!

Farming is a unique industry and one that none of us can do without. The machines and people which produce the food we eat today should be recognised for providing the most important and vital service to the whole of mankind. The next hundred years should be very interesting indeed!

In the past corn would be threshed from the stack when the threshing contractor arrived with his steam-powered traction engine and threshing drum. Here, a Ransomes threshing drum is driven by a belt-pulley from a Field Marshall tractor, ideal for this job as its single-cylinder diesel engine made it economical to run for long periods. ▼

▲ *The arrival of the modern self-propelled combine harvester in the 1940s changed the face of the summer harvest field forever. It replaced the binder, stooking team and threshing contractor with one single machine. Note the operator of this 1970s Massey Ferguson 525 combine is unprotected from the harvest dust.*

SPRING

Top dressing winter-sown wheat and barley is one of the earliest spring jobs. A 93hp four-cylinder Massey Ferguson 3070 is seen spreading nitrogen prills on a crop of wheat, with a fully mounted Sulky DPX 1504 spreader of the twin-disc type that has now replaced the earlier oscillating spout machines. Vanes on the top of the two discs catch the fertiliser as it is fed down from the hopper and then broadcast it in an arc up to 24m wide. An electronic control box is fitted in the tractor cab allowing the driver to adjust both spreading width and application rate from his seat. Nitrogen is applied at this time of year to provide a food source for the young wheat plants and encourage quicker, more uniform growth. ➧

A land-driven Bomford Dyna-Drive preparing a seedbed for sugar beet. At 5m (16.4ft) wide this cultivator needs most of the Lamborghini 190's 189hp, especially as for this job the cultivator is pulled at high speed to break up the ploughed land. The Dyna-Drive is a versatile piece of equipment used equally well as a seedbed former or on minimal cultivation tasks such as stubble breaking. The tines are mounted on two shafts which extend to the full width of the machine and are turned by the force of the tines as they are pulled through the soil, one shaft turning faster than the other. It is available in five widths from 2m to 5m (6.5ft to 16.4ft), and a choice of either packer or crumbler roll is also available for soil consolidation. The Lamborghini has been fitted with all-round dual wheels to spread its weight and reduce soil compaction. ➧

This 125hp six-cylinder Ford 8340 fitted with flotation tyres is also preparing a seedbed for sugar beet, this time with a fully mounted dutch harrow which consists of several rows of spiked teeth behind a levelling bar with a crumbler roll at the rear. Driven at speed, this simple piece of equipment can be an effective way of breaking up clods. Variations of it have been used since the days when the horse was the main form of power on the land.

Sugar beet has been grown in the UK for many years, particularly in East Anglia – the root used to produce sugar while the tops can be used as animal feed. The advent of precision drills saw the demise of the laborious task of thinning the sugar beet crop by hand. The new drills allowed the crop to be spaced accurately and, with the appearance of the monogerm seed which allows only one plant to grow from each individual pellet, sugar beet became an even easier crop to manage. Here, the hoppers on a twelve-row Stanhay Webb precision drill are being filled with seed, each hopper holding up to 6 litres. The tractor is a 100hp six-cylinder Ford 7840 equipped with narrow row-crop wheels which allow the tractor to run in the space between the rows.

Potatoes are grown over a large part of the year. This 107hp German Deutz-Fahr AgroStar 6.08 is using a bedformer prior to potato planting. This bedformer is a pto-driven Howard Rotavator with a Reekie ridger which draws the soil into two ridges. The Rotavator breaks up the soil to reduce clods, a great help to the following de-stoner. Sloping bonnet tractors became popular in the early 1990s with manufacturers such as Massey Ferguson, Landini, Case IH and Steyr following Deutz-Fahr's lead.

A Reekie 330SA de-stoner is shown here at work behind a 100hp four-cylinder John Deere 6400. Removing clods and stones means easier harvesting and a cleaner sample for market. It also helps to reduce damage to harvesting equipment and the crop itself. A share on the de-stoner lifts the ridged soil where rotating paddles guide it on to a series of webs which sift the soil. Stones and clods eventually fall on to a discharge elevator, which can be moved hydraulically from one side of the machine to the other.

▲ *A later, more efficient Reekie de-stoner, fitted with five webs instead of three. A 100hp six-cylinder Ford 7810 draws the Reekie Reliance de-stoner along the rows left by the ridger. A two-wheel drive 85hp Ford 6640 follows with a Reekie two-row planter. The planter cuts two grooves in the smooth bed left by the de-stoner and then drops the seed potatoes into them. Rear-mounted ridging bodies cover the seed and form baulks for the crop to grow into. A levelling bar is attached behind the ridging bodies, leaving the baulks with a level top.*

Cultivator drills have become more and more popular in recent years and have been getting steadily bigger. This Rapide 800P, from Swedish company Väderstad, is 8m (26ft) wide and uses a large chunk of the Caterpillar Challenger 85C's 355hp. The drill uses tines and press wheels to prepare a seedbed while a hydraulically driven fan takes the seed from the large capacity hopper to the coulters. The Caterpillar Challenger 85C comes from North America and features special rubber tracks which allow a fast road speed and do not damage the road surface. This Caterpillar belongs to an East Anglian farmer-contractor who uses it with the Väderstad to drill vining peas for a Suffolk pea-growing cooperative, covering several thousand acres every year.

As the spring starts to warm up sugar beet drilling gets into full swing. The crop is grown in rows 229mm to 600mm (9in to 24in) apart, depending on tractor wheel-track widths when subsequent crop treatment, such as spraying and hoeing, is undertaken. Spacing between each sugar beet plant along the row can also be varied depending on tuber size. An 80hp four-cylinder Massey Ferguson 390 is shown here with a six-row precision drill made by Stanhay Webb of Newmarket, Suffolk. Note the side-mounted guide tines which leave a mark in the soil ready for the driver to line up with on the next pass. This operator is making good use of the tractor's front weight tray to hold extra boxes of seed.

Not long after drilling pre-emergence herbicide is applied to the sugar beet field. This small, low ground pressure Willmot Lightspray 1048 self-propelled sprayer is fitted with wide tyres to prevent soil compaction. Narrower, higher wheels are fitted for work later in the season to allow the sprayer to run between the crop avoiding plant damage. The booms are 12m (39ft) wide and have to be folded and unfolded by hand. Early treatment helps reduce later weed problems and makes subsequent tractor hoeing easier.

This 2500 litre capacity Landquip trailed sprayer, behind a 100hp six-cylinder Massey Ferguson 3085, is shown working in a crop of wheat. Inside the cab an electronic control box allows the operator to adjust the sprayer on the move, making headland work – as shown here, where only one boom is being used to apply chemical – much easier. A pump takes the chemical mix to the 24m (78ft) booms where it is sprayed through nozzles. This example is fitted with optional air sleeve booms which reduce drift by directing the chemical down into the crop, even on windy days. A fan inflates the sleeve with air which is forced out through holes directly on top of the spray nozzles pushing the liquid downwards. Two electronically controlled rams adjust the drawbar, keeping the sprayer precisely within the tramlines. ➧

As the wheat crop grows further doses of nitrogen are applied to speed growth and produce stronger plants. Most fertiliser is now transported in half-tonne 'big bags' like those shown on the forks of this Sanderson 726 materials handler. The Sanderson is filling an Amazone twin-disc fertiliser spreader mounted on the load platform of a Bateman RB15 self-propelled sprayer. ➧

◄ ▼ *Some smaller farms still use hand labour for potato planting. A two-row manned belt-type planter is in use here behind an 82hp Case IH 885XL. Although still needing people on the back to avoid any missed tubers, this type of planter was a definite improvement on planting entirely by hand. On this occasion the crop is being placed straight into a cultivated seedbed without prior ridging or de-stoning; the rear-mounted ridging bodies on the planter form the baulks which will cover the growing potatoes.*

JOHN DEERE
6400

◀ ▲ *As the weather warms up, the sugar beet crop stirs into life. Unfortunately, despite earlier spraying, so do the weeds! Hoeing sugar beet used to be a back-breaking job carried out with a hand hoe, but today a front-mounted tractor hoe does the job. This 100hp John Deere 6400 is equipped with a twelve-row front-mounted hoe. A hydraulic ram lifts the hoe at the headland. Each hoe blade has a spring above it to keep it in contact with the ground. Note that the tractor's wheels are set out so that it can run between the rows of beet. Some farmers like to use a rear-mounted toolbar fitted with two tines which remove wheel marks made when a front-mounted hoe is used, while some prefer to stick to rear-mounted hoes.*

With the approach of summer, farmers who have the facilities start irrigating their crops. This Bauer Rainstar drum irrigator is seen working in a field of winter barley. Water is pumped to the field through underground pipes and a flexible hose is then connected which runs to the irrigator. The water is broadcast in a 48m (157ft) arc over the crop from a special rain gun trolley which is pulled out into the field along the tramlines by a tractor. Water pressure through the drum turns it and pulls the trolley slowly back to the irrigator drum parked out on the headland.

◀ *As the wheat thickens up it needs to be treated against a variety of pests and diseases, such as aphids, wild oats, blackgrass, rust, leaf spot and mildew. The SAM self-propelled sprayer, built by Sands Agricultural Machinery of Stalham, Norfolk, is fitted with 24m (78ft) wide folding booms. Its high ground clearance makes it ideal for following the tramlines through the standing crop.*

Mowers start work on marshes and grassland areas towards the end of May. A six-cylinder Case IH Maxxum 5140 finds its 110hp put to good use with a Claas Corto front and back mower-conditioner combination. The grass is left in a swath to wilt before being collected by a forage harvester or baled for silage. Some farmers leave grass to dry out fully to become hay, but most prefer to turn it into silage for winter feeding of cattle. The Corto mower is of the drum type. Blades at the bottom of each drum cut the grass before it is passed to the conditioning unit which bruises the grass with tines and baffle plates to speed up drying. The grass is discharged from the rear of the machine in a wide swath. ▶

Large self-propelled forage harvesters, such as this 544hp Case IH Mammoth 8790, are most often operated by contractors. The contractor also provides the large number of tractors and trailers needed to keep up with the huge output of these harvesters. The machine's V8 Mercedes-Benz engine is tucked away behind the covers at the rear. A wide pick-up reel gathers in the grass swath which is then fed up underneath the cab to the chopping cylinder. This cuts the grass into smaller pieces, before discharging it at high speed with a high capacity blower into a side-running trailer.

▲ *The last treatment of nitrogen is applied to a crop of wheat. Here a JCB Loadall 525-67 fills a Sulky twin-disc spreader which is mounted on a Massey Ferguson 390. The 106hp JCB Loadall is typical of the many telescopic materials handlers which have become popular with farmers in recent years. This British-built machine features three steering modes: two-wheel steer, four-wheel steer and 'crab', which allows the whole machine to move sideways – particularly handy in constricted spaces.*

▲ *Grass from the forage harvester is carted back to the farm where it is clamped. Many farmers and contractors use materials handlers to fill the clamp and consolidate the grass, though some still use tractor-mounted buckrakes for the task. Large contractors often use high-capacity loading shovels to keep up with the biggest self-propelled foragers. This New Holland LM410 is shown filling a clamp with a very large push-off silage fork. The LM410 is powered by a 106hp four-cylinder turbocharged Iveco diesel engine and has a maximum lift capacity of 2.8t and a maximum lift height of 6m (20ft). A non-turbocharged version of the LM410 is also available with 75hp.*

▲ *Despite the popularity of self-propelled forage harvesters, smaller trailed machines are still common. This JF FCT900 forage harvester from Denmark is typical of modern precision chop machines. It is fitted with a rear-mounted hydraulically operated hitch so that a trailer can be hauled behind it. A 115hp six-cylinder 115hp JCB Fastrac 2115 4WS is shown powering the JF. It is equipped with an electronically controlled four-wheel steering mode to allow tighter headland turns. The trailer is being hauled by the flagship JCB model, the Fastrac 185 which has a 175hp six-cylinder Cummins power unit. The Fastrac concept marries the ability to work in tough ground conditions with a fast road speed.*

◀ The potato planting season is now in full swing and two 125hp six-cylinder Ford 8340 tractors are busy de-stoning land with Reekie machines. The 330SA in the foreground has three webs while the Reliance in the background has five. They are being used by a large vegetable-growing cooperative which uses several of these machines to cover the large area in which it grows potatoes and parsnips. The 8340 was the flagship model in the Ford 40 Series range launched in 1991.

◀ Two potato planters follow the de-stoners. The six-cylinder 100hp Massey Ferguson 6160 and the four-cylinder 100hp John Deere 6400 are both equipped with Keyag planters. Potato planters use either cups or belts to take the seed potatoes from the hopper to the groove in the soil left by a V-shaped share. The tuber is then covered with soil by a set of ridging bodies. If there is a 'miss' on the cup or belt an audible or visible alarm alerts the driver. These Keyag machines are of the belt type.

Matbro from Tetbury, Gloucestershire have made forklifts for many years and entered the agricultural materials handler market early on with their telescopic, pivot steer Teleram. The TS range, which includes this 95hp TS270, was one of the last to carry the Matbro name before production was taken over by the American firm John Deere. This TS270 has a Downs box tipper which uses the handler's hydraulics to grip a seed potato box, cover the open top using a rubber curtain, and tip the box to the left or right. When the box is vertical, the curtain is slid back by means of a hydraulic ram allowing the seed potatoes to fall out at a controlled speed. ➧

As well as grass, livestock farmers also grow maize for silage. A four-row precision drill is shown at work behind a four-cylinder 52hp two-wheel drive International Harvester 454. The drill works in much the same way as the one used for sugar beet, placing the seed in straight rows and at a set distance apart, but it uses air from a pto-driven fan to blow the seed from the hopper to the metering wheel. ➧

◆ *Tedders were first developed for haymaking but their use fell into decline as silage making became more popular. This all changed in the 1990s with the growing interest in grass conditioning. The grass is now mowed and often conditioned on the mower itself. A tedder like this Stoll, seen demonstrated at Grassland '99 on the back of a 120hp Case IH Maxxum MX120, then spreads the swath left by the mower to speed up drying. The same tedder can be used later to row up the grass before forage harvesting or baling.*

◆ *Many farmers are moving away from forage harvesting and clamp making in favour of large bales which are wrapped in plastic to form silage. Claas made their first Quadrant big square baler in 1988. The improved 2200 model was launched in 1999. A 125hp six-cylinder Valtra Valmet 8150 from Finland is shown powering this baler. Hesston pioneered the use of big square balers and were soon followed by Vicon, Massey Ferguson, New Holland, Krone, Welger and Mengele.*

▲ *Potato growers wanting an earlier crop place transparent plastic film over their potato beds before the young plants emerge. The plastic helps retain moisture, heats the soil and prevents frost damage. Once the crop has grown to the point where its leaves are touching the plastic, the film is removed using a machine like the one shown here, mounted on the front of a Deutz-Fahr 6.08. This film remover was built in the farm's own workshop and is fixed to an 'A' frame attached to the tractor's front linkage.*

▲ *As spring rapidly gives way to summer contractors are busy harvesting grass. This team are using a 220hp Claas Jaguar 800 self-propelled forage harvester with a 135hp six-cylinder Massey Ferguson 3635 and a 100hp six-cylinder John Deere 3350, both with Bailey trailers equipped with silage extensions. The Jaguar's 3m (10ft) wide pick-up reel collects the wilted grass from the swath. A metal detector is fitted before the chopping cylinder; this stops the feedroll if a foreign object, such as an implement tine, is found hidden in the grass. This is important as objects that size could damage the harvester's chopping mechanism.*

SUMMER

Potato planting continues into the summer. A Massey Ferguson 3075, with a 98hp four-cylinder Perkins diesel engine, and Reekie 200 de-stoner prepares the land in dry and dusty conditions. The grain bins in the background hold up to 90t of grain and are fitted with a false floor under which air is blown to dry and cool the crop while in store. The Reekie 200 de-stoner is equipped with two webs as well as a scrub web at the rear of the machine which hangs over the last web breaking up any remaining large clods. A pallet has been placed on top of the scrub web to hold it down.

Before the potatoes are planted they are treated with a powder-based chemical which prevents black scurf. The operator is applying the fungicide Monceren, made by Bayer. The planter is a two-row Faun 1400 cup type machine which has been fitted with home-made wooden hopper extensions to increase the holding capacity to just under 1t.

▲ *An 85hp four-cylinder John Deere 2850 conditions swaths of grass left by a mower prior to round baling. This Bamford Wuffler swather was originally designed for hay making and dates from the 1950s. It picks up the crop, bruises the grass – removing the waxiness from the leaves – and then puts it back on the ground in a fluffier swath which speeds up drying. For silage bales a higher dry matter improves palatability for cattle.*

▲ *Round baling is the most popular method of making silage bales. A 150hp six-cylinder Case IH CS150, made in Austria by Steyr (acquired by Case IH in 1996), is seen with a JF Westmac RB 300L round baler which is discharging a bale from its chamber. Here the bales are formed by steel rollers and then tied with baler twine; other balers use rubber belts instead of steel rollers. In addition the size of the bale can be varied. With the growing popularity of silage bale making, baler manufacturers now offer an optional chopping mechanism behind the pick-up reel which chops the grass into fine pieces. This makes denser bales and aids the fermenting process once the bale has been wrapped in plastic.*

▲ *Once the grass has been baled it needs to be wrapped in plastic film. This 100hp six-cylinder Ford 7840 is using a Lawrence Edwards round-bale wrapper which picks up the bale and rotates it on a turntable. The film is dispensed from a holder and, as the bale platform rotates and the bale itself is turned by rollers, the bale is covered in either black or white film. Each bale is wrapped two or three times before being tipped from the rear of the machine, the action of which also severs the film from the dispenser. Several different makes and designs of bale wrapper are available, some designed for large square bales instead of round, some for both and some which can be used in tandem behind a baler – carrying out the baling and wrapping operation in one pass.*

▲ *As the summer draws on, vining peas gradually swell and become ready for harvest. Two Market Rasen Engineering Big Drum viners are shown at work while a John Deere 3350 tractor stands by with a Richard Western high-lift trailer ready to collect the threshed peas. Self-propelled viners first made their appearance in the 1970s, replacing the older trailed machines and pea cutters.*

▲ *Potatoes are among the most intensively irrigated crops because of their high water content and important niche markets which make skin set and colour vital to getting the correct price at market or when meeting contract requirements. Here an Irrifrance Javelin hose-reel irrigator gun trolley is pulled out into the crop by a four-cylinder 52hp International Harvester 454.*

▲ *Although contractors are now commonly used for silage making and, despite the ever-growing popularity of big bale silage, some farmers still prefer to run their own machines. A 100hp six-cylinder Ford 7810 provides the power for this JF forage harvester as it chops the grass and throws it into the side-running trailer hauled by an 85hp four-cylinder Ford 6640. The grass is carted off the marshes and back to the farm where it is clamped.*

▲ Pea harvesters, or viners as they are more often called, have changed greatly since the first self-propelled machines appeared in the 1970s. In 1989, FMC launched their new 979 harvester which had a 300hp Deutz diesel engine, 3.8m pick-up reel and six wheels to spread the machine's 24.6t weight. This FMC 979 is receiving attention from a mechanic after a blockage. With the viners usually working non-stop round the clock the support crew is essential. They replace any worn components as quickly as possible and give the harvesters a complete wash down and service once a day, as well as keeping them supplied with diesel.

▲ *Two main types of mower are used to cut grass for forage harvesting or baling – drum or disc mowers. These more efficient types replaced the much simpler reciprocating knife-bar machines of the 1960s and earlier. This Lely Optimo disc mower is also fitted with a conditioner unit. For road transport it folds up vertically by means of a hydraulic ram. It is shown at work with a 115hp six-cylinder Lamborghini Formula 115 made in Italy by Same.*

◀ ▲ *Wrapped silage bales can be safely handled by hydraulic squeeze forks which grip the bale rather than pierce it. A JCB Loadall 525-4HL materials handler is shown stacking the bales with this type of fork. A Lawrence Edwards stationary three-point-linkage mounted round-bale wrapper is shown at work being driven hydraulically by a 45hp three-cylinder Leyland 245 tractor made in Bathgate, Scotland. When wrapping is complete the bale is ejected from the rear of the machine.*

▲ Many crops need further treatments of chemicals as they grow and start to mature. This crop of wheat is being given its dose of chemical by a SAM Lowline self-propelled sprayer fitted with 24m (80ft) wide booms. Built by Sands Engineering of Norfolk it features a 159hp six-cylinder air-cooled Deutz engine. The forward-mounted cab with its large glass area gives the operator an imposing view out over the crop.

SAM

▲ *An earlier generation of FMC pea harvester is at work here. Power comes from a 302hp eight-cylinder Deutz diesel engine and it is equipped with a 3.3m (10.8ft) pick-up reel. The reel harvests the crop using a 'combing' action which picks up the pea pods but only a small section of the vine, leaving the rest on the ground. Like all pea viners, including the old trailed machines, hydraulic rams keep the machine level at all times, both front to back and side to side.*

◀ *Three FMC 979 pea pickers make short work of a field one summer's evening. Fitted with full hydrostatic drive and with power now increased to 373hp, the 979 can also be converted to harvest broad beans. Despite these huge machines costing in excess of £200,000 each it is not unusual to see them being operated in groups of three, as in this scene, during the pea harvesting season.*

After one crop of potatoes has been lifted the field is ploughed and power harrowed ready for another potato crop. A Lamborghini Formula 115 lines up at the end of the field with a Reekie ridger prior to de-stoning and planting. Note the guides fitted to either side of the ridger making it easier for the operator to keep the ridges straight across the field. This is important as all subsequent work, including planting, will follow the path taken by the ridger. ▶

▲ *The abrasiveness of soil and stones mean that the de-stoner webs and bearings need constant attention. The Keyag two-stage de-stoner is equipped with hydraulic rear-wheel steering to enable tighter headland turns and tracking through the beds, a feature also found on the Reekie and Grimme machines. The Keyag is being powered by a 100hp four-cylinder Lamborghini Premium 1050 equipped with narrow row-crop wheels.*

◀ *A 98hp Ford 7600 from the 1970s follows the de-stoner with a Konings automatic belt-type potato planter. The front of this two-wheel-drive tractor is well ballasted with cast-iron weights which prevent the tractor rearing up on headland turns with a full planter. The 7600 is equipped with Ford's Q cab which became available from 1976 and was the firm's first quiet cab – designed to meet new health and safety regulations. The sticker with the letter E on the side of the bonnet indicates that this is the economy version of the 7600 with a cheaper specification.*

◀ *The next crop ready for harvesting is oilseed rape. This is either desiccated chemically or swathed and left to ripen. For swathing, special machines such as this Shelbourne Reynolds SR2500H are available. These specialist machines are usually operated by contractors. Clearly visible is the 3.6m (12ft) wide cutting header with vertical knives fitted to either side of it allowing the machine to cut cleanly through the crop. The rape is then left in swaths across the field for up to three weeks – depending on weather conditions – before being harvested by combine.*

Potato lifting starts in the summer or even earlier in places such as Cornwall, or where the crop has been grown under plastic. Here a Reekie Cleanflow 2000 trailed harvester is seen harvesting small potatoes for either the canned or pre-packed market. The harvester is unloading on the move. The 1t potato boxes are taken back to the farm and loaded on to large trucks by forklift for road transport to a packing plant where they are sorted, washed and prepared for sale. ▼

▲ *Pea harvesting continues into the summer. Once the peas have been collected from the viners they are loaded on to road transport for their journey to the factory. A 100hp six-cylinder Ford 7840 SLE is shown tipping the contents of its Marston 8t high-lift trailer into a waiting Volvo lorry. Most of the lorries used for this job are owned by farmers who use them as tipper trucks for the rest of the year, the tipper body is removed for the pea season, and a special metal hopper placed on the flatbed.*

When peas were threshed at static viner sites the pea haulm was used for feeding to livestock. Some farmers still use this food supplement, staking out the fields with electric fencing and allowing the animals to feed at will. Other farmers bale the pea haulm from the swath left by the viners after it has been left to dry out, as with this 110hp Ford 8210 Force II and John Deere 550 belt-type round baler. ➧

Injecting slurry, whether animal or human in origin, improves soil fertility. Contractors usually perform this task with large road-going tankers, such as this articulated ERF, which is shown parked on the road next to the field pumping slurry through a pipe in the hedge to a large tanker on the other side. The slurry spreader fills up from the tanker, thereby ensuring a continuous supply. In this case the slurry has been taken from a nearby sewage treatment plant. ➧

▲ *A 144hp six-cylinder John Deere 4255 tractor is shown with a vacuum tanker with injecting tines. The tanker has massive flotation tyres which spread the weight of the machine and reduce soil compaction. The vacuum tanker takes the slurry from the larger tanker on the headland. The slurry is pumped into the large injecting tines which place the slurry below the surface, this reduces smell and avoids surface run-off into any nearby ditches or rivers. In the past, slurry was spread by tankers in wide arcs behind the machine but this is no longer acceptable in most environmentally aware countries.*

As fields are harvested and cleared important maintenance work such as verge and hedge cutting can be carried out before the fields are once again under cultivation. This farmer is taking the opportunity to tidy up around the headland with his 80hp four-cylinder Massey Ferguson 390 and Bomford B487 pto-driven hydraulic hedge cutter. The hedge cutter cuts with flails which revolve at high speed.

It is not long before the fields are put back into use. Here a 185hp six-cylinder John Deere 8100 fitted with very wide low-pressure tyres is seen turning into the furrow at the headland with a seven-furrow Gregoire Besson semi-mounted plough. The tractor's cab-pillar mounted exhaust helps the driver's forward visibility while large areas of glass allow him a good view of the French-built plough.

▲ *At the height of summer comes the high point of the cereal farmer's year — the harvest. The combine harvesters have been standing idle for nearly a year but as the first winter-sown barley ripens they are made ready and, as soon as the weather permits, they suddenly burst into activity. Here in England a church forms the backdrop to a typical modern-day harvest scene, as a 265hp Massey Ferguson 40RS combine chomps through a crop of winter barley while a 124hp Deutz Fahr Commander waits with a trailer to ferry the grain back to the farm. The Massey Ferguson, made in the Dronningborg plant in Denmark, was the biggest in the range at the time.*

As the combines move into the winter barley crop, field after field becomes straw and stubble. The combines need to be moved and the width of modern cutting tables provides a transport problem. This Class Mega 208 is seen lowering its header onto a special trolley, here being towed into place by a Land Rover Defender 90. The whole outfit can then be safely and easily moved to its next destination.

A 165hp six-cylinder Lamborghini Racing 165 drills turnips into virgin winter barley stubble with a Ferrag Rau 3m (10ft) wide combination rotavator and pneumatic seeder. The pto-driven rotavator breaks up the stubble and mixes the soil ahead of the drill's seed tubes. A line of zero-pressure tyres firm the seed into the ground. This drill outfit can be used on ploughed ground as effectively as on stubble. The turnips are being grown for winter feed for cattle and form a cash crop between the winter barley and the sugar beet which will be drilled in the spring.

The latest pea harvester from FMC is the 979-AT Advanced Technology viner which incorporates over thirty design improvements on the previous 979 model. These include a wider podder belt, more controlled aprons to remove trash and an improved 389hp water-cooled V8 Deutz diesel engine built, like all modern tractor and harvester engines, to meet EU emission limits. The unloading hopper has been altered to allow a greater carrying capacity. This harvester is shown unloading on the move into a Richard Western high-lift trailer hauled by a 100hp New Holland 7840.

A 115hp New Holland Ford 8260, built in Basildon, Essex, turns on the headland with a large rigid-tine cultivator. The solid wheels at the rear of the implement act as depth controllers. These machines, often known as chisel plows, were very popular in the late 70s and early 80s when minimal cultivations were much in vogue. Used instead of the plough to rip up the soil, they are also used for deep cultivation work on land destined for root crops such as potatoes. Modern non-ploughing techniques more often make use of single-pass equipment such as the Rau, featured earlier, and the massive drills used in North America (and increasingly in the UK and Europe) such as the John Deere No-Till drill.

▲ *Two large John Deere combines harvest swathed oilseed rape which has been left to dry. The 7.6m (25ft) wide headers of these two machines pick up two rows of swathed rape at a time whilst the rear-mounted integral straw chopper makes short work of the remaining crop residue. The leading combine is a 270hp 2066 while the following machine is a later uprated version, the 2266.*

◀ *Probably the most popular method of harvesting oilseed rape is direct harvesting chemically desiccated crop. Sometimes, if the weather permits, the rape can be left to dry out on its own, but this is not often the case and many farmers find that the results are not as satisfactory. A 235hp Claas Dominator Mega 208 is shown harvesting this tall crop while the close-up shows the electrically driven vertical knife fitted to one side of the 5m (17ft) wide header of a smaller 160hp Claas Dominator Maxi 98SL combine. Claas have been making combines in Germany for many years, producing their first trailed model in 1936. The first Claas self-propelled combine appeared in 1953 with the launch of the SF model.*

◀

Winter barley harvesting is now well under way. A Massey Ferguson 750 combine from the 1970s – and still in regular use – is shown emptying on the move into a trailer hauled by an 82hp four-cylinder Zetor 8111. Cereal harvesting is always a dusty business but today workers are given much more protection than their forebears. The Massey Ferguson 700 Series came with air-conditioned cabs as standard, one of the first combines to do so. The tractor driver also benefits from a quiet, dust-free cab. ➧

Once the barley has been collected from the combine, which has separated the ears from the straw and the grain from the chaff, it is carted back to the farm and usually dried before storage. A 98hp four-cylinder Ford 7610 Force II tips a trailer load of barley into a pit where a cup elevator takes it into the drier to get the moisture level down to 15% or less. After drying, the grain is taken by overhead conveyers into storage bins to await later collection by truck. Barley is used in beer making and animal feed. ➧

Barley straw is much in demand for the feeding and bedding of livestock. A 95hp four-cylinder Ford 7740 SLE is seen baling winter barley straw with a German Krone KR160 round baler. A 2m (6.5ft) pick-up reel lifts the straw from the swath; metal rollers and a moving floor then form the bale which is tied with either plastic twine or net wrap. A JCB Loadall Farm Special 525-67 is shown unloading round bales brought to the farm by a 92hp five-cylinder Lamborghini 955DT and articulated flatbed trailer. The bales are stacked four high using the JCB's telescopic boom, and then left until they are needed for a herd of Friesian dairy cattle during the winter months.

To ensure adequate drainage farmers regularly mole drain their fields. A mole drainer uses a blade with a torpedo at the bottom which cuts a groove through the soil just below subsoil level. An expander towed behind the torpedo leaves a larger tunnel behind the implement allowing water to run into the nearest perforated drainage pipe which then carries the water to field-side ditches. This Miles mole drainer is fitted to a beam mounted on two wheels which is lowered and raised hydraulically. This machine is shown at work in a field of harvested oilseed rape with a 185hp six-cylinder John Deere 8100.

Oilseed rape stubble goes under the plough and the seagulls enjoy a feast as two Case IH tractors turn over the soil. A 190hp six-cylinder American-built Magnum 7220 Pro pulls an eight-furrow Kverneland semi-mounted plough made in Norway, while a 150hp six-cylinder CS150, made in Austria by Steyr, ploughs up the headland with a fully mounted five-furrow British-built Dowdeswell.

◀ *A combine operator changes the knives on a Claas Dominator Mega 208 while a 155hp six-cylinder JCB Fastrac 155-65 keeps close attendance with a large trailer. The British-built Fastrac is fitted with a full suspension system and air braking, giving it fast road speed and making it the ideal tool for grain carting. A load platform behind the cab allows demount sprayers to be fitted, increasing still further the versatility of this equal-size-wheel tractor.*

Stubble is broken with a Kongsklide Vibroflex cultivator which is fitted with thirteen tines. It is seen on a 115hp six-cylinder British-built Ford 8200 first launched in 1980. The cultivator is fitted to the tractor's three-point linkage allowing for quick headland turns and easy road transport. ▶

▲ *Winter barley stubble has been cultivated before turnips are drilled with an impressive looking Accord drill. This drill has a front-mounted hopper and Maschio 6m (20ft) wide power harrow, all mounted on a 165hp six-cylinder Lamborghini Racing 165. The Maschio power harrow is made in Italy and, like the drill, folds up hydraulically for road transport. A fan under the front hopper blows the seed through the pipe along the tractor and then into the drill coulters at the rear. Note the two discs at the very rear of the outfit, these mark out the tramlines for later pre-emergence treatment.*

After drilling, the turnip seed is rolled into its rough seedbed with a set of hydraulically folding rolls hauled by a 100hp four-cylinder John Deere 6400. Rolls like these are used to form seedbeds before drilling as well as to roll established cereal crops in the early spring.

The McConnel bale packer appeared in the 1970s and was in many ways the predecessor of the modern big baler. It hitches behind a conventional square baler – in this case an International Harvester 440 – and packs twenty rectangular bales into one neat package. This means that a suitably equipped loader can then cart the packs off the field in bulk. This is more efficient than the sledge which only collected eight bales at a time. The Case IH Maxxum 5130 Plus, with 100hp and six cylinders under its bonnet, has plenty of power to spare on this job.

INTERNATIONAL
CASE III
5130
PLUS

The Case IH 527 combine is made by MDW in eastern Germany. The American company Case sells it alongside its rotary Axial Flow combine range. This 527 is shown here coping with a lodged crop of winter barley while on demonstration, having just arrived in the UK. Header width is 7.2m (23.6ft) and power comes from a six-cylinder 271hp diesel engine.

A Valmet 8100 bales barley straw with a Welger RP200 round baler. The bales are being net wrapped, with spare wrap stored on the front linkage. Valmet tractors are made in Finland and Welger are now owned by Danish firm Lely. The harvest field scene has gradually changed from the small rectangular bales, which dominated for over 30 years, to large round bales and the latest fashion – big square bales.

Same-Deutz Fahr combines are built in Germany and the larger models are badged Topliners. These combines shown harvesting winter barley are 4075HTS models featuring a 240hp six-cylinder air-cooled Deutz engine and full hydrostatic drive. These two machines are operated by a contractor and, with several thousand acres to cover, they often work into the night, moisture content permitting.

PLOEGER
PLOEGER
PLOEGER

By August the pea season is coming to an end. Ploeger, from Holland, build pea harvesters like this EPD 520 model which features the Vista cab used by Claas. It is powered by an eight-cylinder air-cooled 295hp Deutz engine and uses a different type of unloading system to the other viners featured in this book – a moving conveyor belt folds down from behind the cab to unload the peas into a side-running trailer. This viner has a different layout to the FMC machines. The cab is positioned well forward over the header and the engine is mounted at the rear. Something that it does share with all pea viners is its hydraulically operated all-round self-levelling. This keeps the machine level even on hilly ground so that the threshing drum maintains efficiency.

▲ *With 170hp on tap, this Massey Ferguson 3670 copes easily with a fully mounted Dowdeswell five-furrow plough while turning in oilseed rape stubble. The yellow tool attached to the plough is a Claydon furrow cracker; it is equipped with a row of tines which shatter the clods as it is drawn through them. It is attached to the plough frame and turns over with it on the headland ready for the next pass. All the ploughs shown in this book are of the reversible type which have two sets of bodies, allowing the driver to start at one side of the field and then work up and down until he reaches the other side. The smaller non-reversible ploughs used in the past needed the field to be expertly marked out before work could begin. The Massey Ferguson 3670 was built in Beauvais, France with a six-cylinder engine made in Finland by Valmet.*

◀ *It is August and the wheat harvest is well under way. A 235hp Claas Dominator Mega 208 prepares to empty on the move into a Richard Western 12t trailer hauled by a 93hp Massey Ferguson 3070. The straw is being chopped so that it can be ploughed back in to improve soil fertility – a much better way of dealing with it than the now banned practice of straw and stubble burning.*

◀ *When the harvested wheat arrives back at the farm it is tipped from the trailer into a grain pit and fed into a drier before storage. This farm needs a large pit to cope with the output of its Massey Ferguson 38RS combine. The tractor is a 125hp six-cylinder Lamborghini 1306; the trailer was made on the farm from an old Leyland tipper lorry.*

While the main part of this spring barley field has been round baled, the headland is being square baled by a large Hesston 4755 square baler complete with a sledge which collects two of these huge bales together. The tractor is a Case 2090 built in America and the bales are destined for use by a large pig farming enterprise.

Although wheat straw is not as highly sought after as barley straw it is still a good bedding material for cattle and is often used. This JCB Loadall 525B-4 materials handler is loading a trailer with round bales ready for carting back to the farm. Machines such as the Loadall were originally designed for work on construction sites but it did not take long before their usefulness in agriculture was realised. Their four-wheel drive, telescopic boom and manoeuvrability make them ideal tools for both livestock and arable farmers. By the 1990s many different manufacturers were offering complete ranges of these machines designed specifically as agricultural materials handlers.

▲ *Meanwhile the wheat harvest continues, as this Case IH 2166 Axial Flow reaches the headland. Powered by a 240hp engine, the 2166 uses a rotary threshing system as opposed to the conventional threshing drum and straw walkers. The straw chopper kicks up some dust, helped by the rear-mounted chaff spreaders. More and more farmers, particularly in the USA, are now using yield mapping systems on their combines. Most of the main combine manufacturers now offer factory-fitted global positioning systems (GPS) which use satellites to pin-point the combine's exact position in the field. This data is combined with information on the amount of crop harvested from that part of the field to form detailed and accurate yield maps.*

▲ *Modern heavy machinery can leave its mark once it has left the field. Soil compaction can have a major impact on the cereal grower's resulting yields so, as the cereal crops are cleared many farmers, especially those on heavy land, make a pass across the fields with a subsoiler. This works like a cultivator with three or more legs designed to cut through the soil as they lift it, breaking up the pan. This 186hp six-cylinder Ford TW35 Force II, fitted with the Super Q cab launched in 1985, is using a Flatlift three-leg subsoiler to break up the tramlines, often the most compacted area of the field.*

▲ *The clods of soil shine in the summer sunshine as a 110hp six-cylinder Case IH Maxxum 5140 Plus turns over the heavy clay soil with a five-furrow Gregoire Besson fully mounted plough. In this case the plough is set at 40cm (16in) furrow widths but, by moving the angle of the plough bodies either away from or closer to the main beam, the furrow width can be adjusted from 30cm (12in) to 50cm (20in). Some ploughs feature hydraulically operated change-on-the-move furrow width adjustment.*

▲ *The combine harvester is used to harvest many different crops from sunflowers to maize. A Claas Lexion 460, with 300hp at its disposal, makes short work of this field of broad beans with its huge 9m (30ft) wide header. The beans have been left to blacken and thoroughly dry out before harvesting, making it easier for the combine to thresh them from their pods. A 130hp six-cylinder John Deere 6900 collects the beans in a Richard Western trailer before carting the crop back to the farm. Broad bean harvesting will be the last job for this combine as the summer draws to a close.*

AUTUMN

◆ ▶ Autumn is one of the busiest times of the year on the farm. Here lime is applied to the fields to reduce acidity in the soil. Contractors are usually used for this job and this one covers a lot of ground with his British-built JCB Fastrac 155-65 and KRM Bredal lime spreader fitted with large flotation tyres. A French-built 85hp Massey Ferguson 6130 is used to load the spreader with its Massey Ferguson 905 front loader. Lime spreading is becoming more precise with the advent of GPS technology. Yield maps from the combine can be used to highlight particular areas of the field which need treatment.

GRANDAD 2
FASTRAC
KEITH
MOUNT

◀ ▼ *Not all ploughs are the same, or use the same method to work – the first ploughs were of the one-way type, now reversible ploughs are used. In some countries the disc plough is popular and some farmers use conventional ploughs with slatted mouldboards. In the early 1990s the plough story took a new twist with the appearance of the square 'plow'. This Opico four-furrow is at work behind a 110hp Case IH Maxxum 5140 Plus. The mouldboards are square and made of plastic, the share and wings are the only soil-working components made out of metal. To allow ploughing up and down the field like a conventional reversible the main beam of the Opico swings from left to right so that the mouldboards always turn the soil in the same direction.*

▲ *Mole draining is often carried out by large-wheeled tractors, but some farmers still prefer the traditional form of power – the steel-tracked crawler tractor. The Track Marshall TM135 is powered by a 135hp Perkins six-cylinder diesel engine and is equipped with full-length steel tracks. It was built in Gainsborough, Lincolnshire. The picture shows the crawler turning on the headland with a Miles mole drainer in raised position. Older crawlers such as this are not fitted with a steering wheel but the brakes are applied to one track at a time, enabling the tractor to slew round.*

Oilseed rape stalks are incorporated prior to ploughing here with a six-cylinder 130hp John Deere 6900 and set of offset disc harrows. This is the sort of job that disc harrows are most often used for today, but they were once very popular for breaking down ploughed land to form a seedbed, a task now more often done by the power harrow. ▶

A Dowdeswell fully-mounted seven-furrow plough complete with double-ring press at work in sticky clay soil. Power comes from a 200hp Track Marshall TM200 rubber-tracked crawler. Most ploughs of this size are semi- rather than fully mounted, and this extremely heavy piece of equipment managed to bring even the well-ballasted Track Marshall's nose up when lifted on the headland! The Track Marshall's rubber tracks mean that it is capable of road travel. Unfortunately the TM200 did not prove as popular as the original rubber-tracked model, the American Caterpillar Challenger 65, and it was only in production for a few years.

◆ *Power harrowing with a Lely Roterra on a six-cylinder 100hp Ford 7810. Instead of duals, the tractor has been fitted with metal cage wheels to help spread its weight over a greater area. Lugs around the inside of the tractor wheel centres connect with lugs on the cage wheel, a threaded rod is used to pull the two sets of lugs together. Although once very common, cage wheels are now quite a rare sight as many farmers prefer dual wheels or single flotation tyres.*

◆ *A four-cylinder 85hp John Deere 2850 drills winter barley with a 4m (13ft) wide Juko drill. The Juko's tramline markers are operated by a small electric motor positioned on the right-hand side of the drill. Winter barley is usually the first autumn-sown crop, quickly followed by wheat.*

A four-cylinder 95hp Ford 7740 SLE turns in green stubble with a four-furrow Ransomes plough made in Ipswich, Suffolk. Many farm workers still take great pride in their ploughing skills. The ever-popular National and International Ploughing Championships and the large number of smaller ploughing events up and down the country bear testament to the skill of the modern ploughman, even in the rush of modern-day farming.

After ploughing, the land is rolled to break down clods and retain moisture. A set of Väderstad 6m (20ft) wide hydraulically folding rolls are seen at work behind an 82hp four-cylinder Case IH 895XL DUO built at Doncaster and equipped with all-round dual wheels.

▲ *This American-built 185hp John Deere 8100 with large flotation tyres is sowing wheat at speed with a Väderstad Rapide 400P 4m (13ft) wide cultivator drill. The Vaderstad is drilling into a seedbed prepared in just one pass by a trailed double-ring press; this type of drill does not need a lot of soil preparation ahead of it. A fan on the drill transfers the seed from the hopper to the coulters and can be driven by either the pto shaft or the external hydraulics.*

◀ *Large-scale farming requires large-scale machinery. This contractor's outfit consists of a 325hp six-cylinder Caterpillar Challenger 75C crawler and an eleven-furrow articulated Kverneland semi-mounted plough. The Caterpillar, a road-going rubber-tracked crawler, is more versatile than its steel-tracked predecessor which had to be carted from field to field on a low loader. The plough's articulated beam is divided into two, with seven furrows on the first section and four on the rear. All adjustments are controlled from the cab. Norwegian firm Kverneland introduced their first tractor plough in 1930, with a reversible model appearing in 1973. The plough shown here is part of the PX/RX range made near Stavanger in Norway. The PX ploughs feature shear-bolt protection whilst the RX versions are fitted with Vibromat shock absorbers.*

This Simba trailed double-ring press is used to form a seedbed straight after ploughing. It uses a set of rigid tines to break up the clods, steel rings then follow to firm the soil and cut through any remaining clods. For transport the implement is pulled from the drawbar on one end. A set of hydraulically operated wheels fitted to the opposite end raise the whole machine off the ground. Although the power harrow is still the most often used cultivation tool, equipment such as this Simba press are becoming increasingly popular with many growers wishing to cut crop establishment costs as well as limit soil compaction. ▶

▲ *The Ransomes Nordsten drill is a good example of what is commonly known as a corn drill. It is seen here sowing wheat behind a two-wheel-drive, four-cylinder 98hp Ford 7710 Force II equipped with dual rear wheels. This 4m (13ft) wide drill is fully mounted on the tractor's three-point linkage and is land-wheel driven.*

▲ *The Standen pulverisor mounted on the front of this 100hp John Deere 6400 makes short work of the potato leaf, leaving clean beds for the Reekie CleanFlow 2000 two-row potato harvester following behind. The pulverisor removes the tops with a series of rubber flaps which revolve at high speed, each flap is set at a different height to allow for the unevenness of the potato baulks. The tops are discharged to the right of the machine and the whole outfit is operated by hydraulic motors. The harvester lifts two rows at a time, the crop is then carried over a series of webs to a roller table where any trash and stones are removed by rubber or steel rollers, depending on ground conditions. The crop moves on to the picking-off table at the rear. This can accommodate up to four people who remove anything other than potatoes – including any bad tubers – which get past the rollers. From the picking-off table the potatoes are taken to an unloading elevator which can be raised or lowered hydraulically.*

A large North American tractor turns over the stubble. This 310hp Case IH Steiger 9350 Row Crop Special comes complete with massive flotation tyres and a Naud ten-furrow semi-mounted articulated plough from France. Steiger started making large, four-wheel-drive tractors in Minnesota in 1957 and were taken over by Case IH in 1986.

The latest Caterpillar rubber-tracked crawlers are now sold in Claas colours in Europe. This Claas Challenger 65E with its 310hp six-cylinder CAT engine is shown at work preparing a seedbed for oilseed rape with a 6m (20ft) wide Kuhn power harrow. Rubber-tracked machines such as the 65E reduce soil compaction – the large surface area of each track means that the weight of the machine is spread over a much greater area than an equivalent wheeled tractor.

⬆ *Filling the drill the easy way. Wheat seed is tipped from a Richard Western dumper trailer into the waiting hopper of a Massey Ferguson 510 pneumatic drill made for Massey Ferguson by Greenland. The drill has a working width of 6m (20ft) and both outer sections fold up for road transport. The drill can be controlled from a cab-mounted control box.*

▲ *Sugar beet harvesting begins as early as September. A large six-row self-propelled sugar beet harvester unloads on the move into a Richard Western 12t dumper trailer hauled by a 93hp four-cylinder Massey Ferguson 3070. The Riecam RBM 230 is fitted with full hydrostatic drive, a 2t holding tank and a 230hp Deutz engine. A potato harvester version was also available using the same chassis and running gear.*

◀ ▶ Today, sugar beet harvesting which used to be so labour intensive is often undertaken in a single pass. A six-row French Matrot harvester is shown at work here wrapped around a Deutz-Fahr Agrostar 6.08 from Germany. A front-mounted topper uses knives to remove the leaves from the sugar beet crowns and an auger discharges them to the side. The roots are then lifted by shares on to a set of cleaning turbines where excess dirt is removed. An elevator carries the crop up to the discharge elevator and into a side-running trailer, in this case hauled by a 98hp four-cylinder Ford 7610 Force II.

Sugar beet land is turned in by a 145hp six-cylinder German-built Case IH 1455XL and Lemken six-furrow fully mounted plough. Front weights are attached to the tractor's front linkage to balance the considerable weight of the plough whilst turning on the headland. ▶

▲ *Harvesting potatoes with a self-propelled Amac ZM2 two-row harvester. The tops have been killed off with sulphuric acid and this crop is being harvested to be used for seed the following year. Despite their popularity on the continent and in America, not many self-propelled potato harvesters are to be seen at work in the UK.*

▲ *A New Holland 8260 at work with a Dowdeswell five-furrow plough and trailed furrow press. The press is towed behind the plough by a drawbar which disconnects from a special press arm on the plough when it is lifted at the end of the field.*

▲ *This picture clearly shows the difference between ploughed land and that left after a pass with the power harrow. The 85hp four-cylinder Ford 6640 SL is using a 3m (10ft) wide Kuhn power harrow, made in France, as it follows the plough on a sunny autumn day.*

▲ ▶ *A 61hp three-cylinder Ford 4610 drills wheat with a 4m (13ft) wide Juko drill. A Case forklift is used to bring the seed to the field, the drill is then filled by hand. The farmer is shown adding slug pellets to the hopper which are sown with the seed. Some farmers apply slug pellets to the growing crop later using a small broadcaster mounted on an ATV (All Terrain Vehicle) such as a quad bike.*

▲ *A 135hp six-cylinder New Holland 8360 equipped with Standen row-crop wheels pulls a Standen Status 1750 two-row potato harvester, while a 120hp six-cylinder Massey Ferguson 6180 drives alongside with a trailer. Standen are based in Ely and manufacture a large range of potato-related machinery. The Status is just one of the several harvesters they produce.*

▲ *Forage maize harvesting is most often carried out by a contractor. Here a 320hp Mengele Mammut 6300 self-propelled forage harvester complete with five-row Kemper rotary header is attended by two Italian Same Antares 110 110hp six-cylinder tractors and trailers equipped with silage extensions. The chopped maize is carted back to the farm where it is clamped and made into silage in much the same way as grass.*

CASE INTERNATIONAL
case
885XL

◀ ▶ ▼ *Potato harvesting with a single-row Grimme Row Over from Holland. An 82hp four-cylinder Case IH 885XL pulls the Grimme while a little 45hp Massey Ferguson 135 pulls a trailer alongside. Inside the harvester, pickers are employed to remove any haulm, stones or damaged tubers. A series of webs carry the crop through the harvester until the potatoes reach the unloading elevator, the height of which can be adjusted from the tractor cab. It is generally kept as low as possible to reduce damage to the potatoes. The other harvesters featured use rollers to remove soil from the crop but the Grimme uses a pintle belt for this job. This consists of a rubber belt with raised fingers which hold onto the soil as potatoes hit a raised plate which deflects the crop on to the picking-off table, the soil eventually falling back down on to the field.*

Turkey muck is a very useful fertiliser. Here contractors spread it on to stubble using two ECE Fieldmaster 1200 rear-discharge spreaders. These machines spread up to 12m (40ft) wide using a moving floor and rear-mounted spreading discs. A Ford A64 industrial-wheeled loader fills the spreaders. Two generations of Renault tractor from France are in charge of the spreaders. The older, a Renault 155-54 powered by a 145hp six-cylinder MWM engine and complete with fully suspended TZ cab, is shown with its spreader being filled. The tractor seen spreading is a newer 145hp Renault Ares 710 RZ – the replacement for the 155-45 – with power from a six-cylinder Deere Power Systems engine. The new Ares tractors also feature a suspended cab which gives the driver a much smoother ride, even over rough ground.

PAUL GUNTHER
-CONTRACTING-
01603 867150
ECE
RENAULT
710RZ
Ares

As the autumn progresses sugar beet harvesting becomes a more common sight. Here a 95hp six-cylinder Case IH 956XL carts sugar beet away from the harvester in a Richard Western dumper trailer. The six-row self-propelled harvester is a 275hp Moreau GR4 MKII made in France. The tops are removed by the front-mounted topper; flails take off the majority of the leaf and knives then remove the top of the root. Lifting shares then take the roots to five turbines where the beet is cleaned and then passed under the cab and engine to the elevator at the rear which carries it into the holding tank. ➧

➧ *Not all contractors use big self-propelled forage harvesters for the maize crop – some use a tractor and trailed forage harvester. However, the tractor-drawn harvester cannot match the self-propelled machine, the biggest of which can harvest up to eight rows of maize at a time. This contractor has found a compromise; a Steyr reverse-drive tractor from Austria when fitted with a Kemper maize header and forage harvester can certainly match the output of smaller self-propelled machines. A 110hp six-cylinder Fiat 110-90 and 85hp four-cylinder Case IH 856XL cart the chopped maize away with two trailers, both equipped with silage extensions.*

Sugar beet harvesting with another French machine, this time a Matrot M31 Electronic six-row self-propelled harvester powered by a 237hp Deutz eight-cylinder diesel engine. Two 120hp 6800 John Deere tractors cart the crop away, one with a Richard Western dumper trailer the other with a Brian Legg dumper. The Matrot has been fitted with dual rear wheels on the right-hand side.

Once lifted from the ground, the sugar beet is either taken back to the farm or clamped out in the field to await transport to the processing factory. An 82hp four-cylinder Case IH 895XL DUO built in Doncaster waits next to a field clamp with a homemade 'pusher' tool attached to its Tanco front-end loader. This device is used to push up the pile of sugar beet and form a level top to the clamp, preventing frost forming in any undulations which could cause the roots to rot.

▲ *Two tractors from Massey Ferguson harvest potatoes for the pre-pack market. A 110hp six-cylinder 6170 pulls the Reekie Cleanflow 2000 two-row harvester and front-mounted Standen pulverisor – not needed on this field as the tops have been killed with sulphuric acid. An older 100hp six-cylinder 3085 drives alongside the harvester collecting the crop in potato boxes, each box containing 1t of potatoes, on a Brian Legg flatbed trailer capable of holding eight boxes.*

Most fields need proper drainage if maximum yield is to be obtained. Contractors are usually employed for this task. This contractor is using a Mastenbroek self-propelled drainage machine, fitted with large steel tracks. One of the marker posts used to guide the machine across the field in a straight line – by means of a laser mounted both on the drainage machine and on marker posts – is shown in the foreground. A large blade cuts a groove through the soil leaving a trench into which plastic drainage pipe is layed before backfilling with stones and soil. ▶

◀ *The latest concept in sugar beet harvesting is the tanker harvester. These huge machines are built to a completely different specification to the self-propelled harvesters shown previously – all the topping and lifting gear are ahead of the front wheels which allows the harvester to be fitted with four large flotation tyres. A large capacity tank forms the middle of the machine, with the cab in front and the engine mounted at the rear. This Verveat sugar beet harvester is a good example with power coming from a 330hp turbocharged DAF diesel engine and a holding tank with a capacity of up to 17t.*

Once the sugar beet has been carted away from the field and clamped, it then begins its final journey. Cleaner loaders, like this CTM Euro 180, are used to load the roots into a lorry for the trip to the processing factory. A loader is used to fill the hopper, the beet are then cleaned by metal traces above and below the crop as they travel up to the unloading elevator. The unwanted soil falls through beneath the machine and is usually returned to the land.

Cereal crops are growing well by the time winter arrives and will need chemical treatment before the arrival of the really harsh weather when they will become dormant for a while. Winter barley is being treated here by a 95hp four-cylinder Ford 7740 SLE and Lely 12m (39ft) wide fully mounted sprayer. Farm sprayers become very busy in the late autumn and early winter periods, applying a wide variety of different chemicals.

WINTER

◂ ▾ *Although the majority of the sugar beet making its way to the processing factory has been lifted by contractors with large, six-row machines, many farmers still prefer to harvest their crop themselves. Often they use older small machines like this single-row Standen Rapide MK2, which tops the crop a row ahead of the lifting gear and collects the beet in a hopper so that a trailer does not have to run continuously beside it across the field – particularly useful when the ground is wet.*

▲ *A 110hp John Deere 6600 tractor draws away from the sugar beet clamp with a Richard Western dumper trailer in raised position after adding another 10t of beet to the clamp. In the background a JCB Loadall 525-67 materials handler brings two round straw bales to cover the clamp to prevent frost damage. Some farmers use sheets specially made for the job. Care has to be taken however to ensure that the crop does not overheat whilst covered as this could cause as much damage as the frost itself.*

▲ *As the sugar beet fields are cleared, tractors move in to prepare the ground for the following crop. Here a 170hp six-cylinder John Deere 7800 tractor built in America turns over the soil with a French Gregoire Besson eight-furrow semi-mounted articulated plough. A 120hp John Deere 6800 follows closely behind with a 4m (13ft) wide Maschio power harrow complete with rear crumbler bar. In the background is a 130hp six-cylinder John Deere 6900 drilling wheat with a 4m (13ft) wide Simba Freeflow cultivator drill. The 6900 is equipped with dual wheels all round while the 6800 is fitted with dual rear wheels only.*

Front linkages on tractors have become increasingly popular in recent years. Most tractor makers offer factory-fitted front linkages for their machines, but there are also plenty of retrofit linkages on the market. This 125hp six-cylinder Lamborghini 1306 is drilling wheat with a 3m (10ft) wide Ferrag Rau rotavator drill, while its front linkage is being used to push a single-ring press which breaks down the clods between the tractor wheels.

This 110hp six-cylinder John Deere has a Lynx Sumo zero-pressure tyre press mounted on its front linkage while drilling wheat and is also equipped with all-round dual wheels. It is using another type of combination drill, which was one of the first of its kind to appear. A bridge-link attachment links a 4m (13ft) wide Lely Roterra power harrow with a Bettinson grain drill of the same width. Both implements can be detached and used separately if necessary. The bridge-link pivots so that headland turns can be negotiated.

◆ ◆ *A 100hp six-cylinder Case IH Maxxum 5130 Plus reverses up to the drawbar of a 10t capacity Rolland Turbo 10 trailer made in France. Once connected up, the tractor and trailer are ready to travel beside the sugar beet harvester. Three rows of sugar beet are harvested at a time with the Standen Turbobeet MK3. A topper is mounted on the front of the 95hp four-cylinder Massey Ferguson 3075 which removes the bulk of the leaves, leaving a linkage-mounted scalper to finish the job behind the tractor, ahead of the harvesters opel wheel lifters which squeeze the beet out of the soil and place the roots on to the main trace. This then carries the beet to the discharge elevator at the rear of the harvester. A trash web removes any grass or other leafy matter from the trace series of teeth on the trace bars.*

TURBO 10
Standen
Turbobeet
3075

A 102hp four-cylinder JCB Loadall 525-67 finds itself temporarily working on highway maintenance duties as the winter takes hold and the narrow country roads quickly fill with snow. This machine's telescopic boom and large bucket make it ideal for such work. Four-wheel drive and three steering modes mean that they are also suited to a myriad of other tasks. ▶

◀ *Despite the snow and frost, some sugar beet harvesting still needs to be done. This Riecam RBM 230 lifts the first strip around the headland of a field, leaving a prominent line of brown soil behind.*

▲ *The German Holmer Terra Dos self-propelled six-row sugar beet harvester is a harvester with a difference. As with all tanker machines the topping and lifting gear are ahead of the cab and front wheels. A large holding tank is positioned behind the cab, but the difference is that behind the cab the rest of the machine is mounted on a pivot, making the harvester articulated. During harvesting the rear wheels can be run offset to the front wheels, spreading the harvester's weight over a larger area and reducing soil damage. At the headland the rear wheels line-up with the front once again for turning, a task made easier by the machine's articulation. Here a brand new Holmer Terra Dos is shown at work with a 135hp six-cylinder JCB Fastrac 1135 4WS and Bailey trailer in Norfolk. The harvester is controlled electronically from the cab; the operator also uses a touchpad computer monitor to ensure correct functioning.*

◆ ◆ *An 82hp four-cylinder Ford 6610 cleans parsnip beds with a three-point-linkage mounted brush before harvesting. Normally this tool is used to remove any leaves or weeds from the bed. Here, however, it is removing a covering of snow. A Dutch Amac harvester follows the brush, harvesting one bed of parsnips at a time, each bed containing five rows. A 110hp six-cylinder Massey Ferguson 6170 pulls the harvester. The front-mounted pulverisor is not being used but has been left in place after potato lifting to provide extra ballast at the front of the tractor, improving grip in extremely wet and difficult lifting conditions. Two 105hp six-cylinder John Deere 6506 tractors with Richard Western trailers cart the crop away to be washed and graded before sale.*

TOPMASTER

RBM 300 S
RIECAM

◆ ▲ *Another tanker type sugar beet harvester, the Riecam RBM 300 S is powered by an eight-cylinder turbocharged 300hp diesel engine fitted at the rear of the machine. From the Claas cab the driver has a clear view of the four cleaning turbines, which on some types of harvester are positioned out of the operator's view. A single joy stick controls all the main functions. Imported into the UK by Armer Machinery, the Riecam features automatic steering while harvesting and a 14t holding tank. One of the main benefits of the tanker system is that tractor and trailers can be left on the headland of the field and only the harvester needs to travel up and down, preventing damage to the soil in wet conditions. Most of the time, however, emptying is carried out on the move during harvesting. Here a New Holland 8340 and trailer is dwarfed by the Riecam beside it.*

▲ *A 150-head dairy cattle herd provide the manure for this Richard Western rear-discharge manure spreader. An 92hp Lamborghini 955DT applies the warm muck to a field of grazed stubble turnips. A slatted moving floor draws the manure towards a set of rear-mounted beaters which throw the load out behind the machine.*

▲ *Winter ploughing can be tricky if the weather is not on your side. Double Italian muscle is needed here to pull a seven-furrow Gregoire Besson semi-mounted plough through extremely wet stubble. An 115hp six-cylinder Lamborghini Formula gives a helping hand to a 165hp six-cylinder Lamborghini Racing 165, giving a total of 290hp, twelve cylinders and eight wheel drive!*

With the arrival of a new year new plantings begin. A 150hp six-cylinder John Deere 7700 prepares three beds at a time for parsnip planting. This machine follows the beds left by a de-stoner, its rotary paddles level the bed ready for the planter. Preparation has to be got right as these beds may need to be kept intact for nearly a year before harvesting takes place.

The parsnips are drilled with an 82hp four-cylinder Ford 6610 made in the Basildon factory in Essex in the 1980s. It is using a five-row air-operated precision drill planting the seed in the beds left by the bed former. The tractor's front weight tray has been extended and modified to accommodate several boxes of parsnip seed, allowing the operator to fill up the drill with seed whenever necessary.

Spring barley drilling often starts towards the end of winter, providing the soil has warmed up sufficiently for the seed to begin germination. A 110hp six-cylinder Massey Ferguson 3095 uses a 3m (10ft) wide Simba Freeflow cultivator drill in February. When the machine turns on the headland it is raised up on rear-mounted zero-pressure tyres which lift the drill out of work. The large rigid tines which break up the ploughed land can also be clearly seen. British company Simba, founded in 1976, introduced the Freeflow drill in 1994.

As the wheat and barley start to grow they have to be treated with chemicals and fertilisers. A 130hp John Deere 6900 prepares to leave the field it has just completed with a 24m (78ft) wide Chafer trailed sprayer.

Silhouetted against the intensity of the evening sunlight that streaks across the sky,
the monster-like machine moves slowly along the stark outline where clouds and land meet.
Harvesting the tender crop of peas with the gentleness of a snowflake cascading to the ground,
the viner continues to plod relentlessly across the fields, until the pea season is finally complete.

▲ *An FMC 979 pea harvester works into the sunset as the twentieth century draws to a close. In 1900 farming all over the world was carried out by hand labour and horse power; the steam-powered traction engine being the ultimate in farm machinery technology. Internal-combustion-engine tractors were struggling with reliability problems and it would be another twenty years before they really began to be accepted seriously. How could anyone interested in farm machinery in the early 1900s have envisaged the huge six-wheeled, 373hp diesel engined FMC 979? And with technology advancing at its current pace no one can predict where the next hundred years will lead us.*

INDEX